JUPITER

MARYSA STORM

BLACK RABBIT BOOKS

Bolt Jr. is published by Black Rabbit Books
P.O. Box 227, Mankato, Minnesota, 56002.
www.blackrabbitbooks.com

Michael Sellner, designer and photo researcher

Names: Storm, Marysa, author.
Title: Jupiter / by Marysa Storm.
Description: Mankato, Minnesota : Black Rabbit Books, [2023]
Series: Bolt Jr. space explorer | Includes bibliographical references and index. | Audience: Ages 6-8 | Audience: Grades K-1 | Summary: "Suit up! Jupiter is a planet like no other, and it's time to explore it through a colorful design that features vibrant photography, engaging infographics, and carefully leveled text"– Provided by publisher.
Identifiers: LCCN 2021021506 (print) | LCCN 2021021507 (ebook) | ISBN 9781623107734 (hardcover) | ISBN 9781644665947 (paperback) | ISBN 9781623107819 (ebook)
Subjects: LCSH: Jupiter (Planet)–Juvenile literature.
Classification: LCC QB661 .S76 2023 (print) | LCC QB661 (ebook) | DDC 523.45–dc23
LC record available at https://lccn.loc.gov/2021021506
LC ebook record available at https://lccn.loc.gov/2021021507

Image Credits

Alamy: Artur Maltsau, 16–17; Stocktrek Images, Inc., 4–5; Dreamstime: Planetfelicity, 18–19; iStock: Elen11, 12–13; Science Source: Detlev van Ravenswaay, 1, 10–11; GREGOIRE CIRADE, 10–11; NASA/JOL-Caltech/SwRI/JunoCam, 22–23; Victor Habbick Visions, 6–7; Shutterstock: 42videography, 20–21; Artsiom P, 4–5; Bikomins, 3, 24; Mopic, 14–15; RHJPhtotoandilustration, 20–21; Vadim Sadovski, Cover; Vectorfair.com, 14–15; Vector Tradition, 12–13

Contents

CHAPTER 1

Meet the Planet

Deep in space, a giant storm **rages** on Jupiter. The storm looks red. It's been swirling for hundreds of years. People call it the Great Red Spot.

rage: to continue out of control

COMPARING SIZES

Jupiter

272,946 miles (439,264 kilometers) around

A Big Planet

Like the storm, Jupiter is huge. In fact, it's the solar system's largest planet. It's about 11 times bigger than Earth.

Features

bands of clouds
four sets of rings
smaller storms

CHAPTER 2

Its Place in Space

Jupiter is made mostly of **gas**. It has a small **core**. The planet's rings are rock and ice.

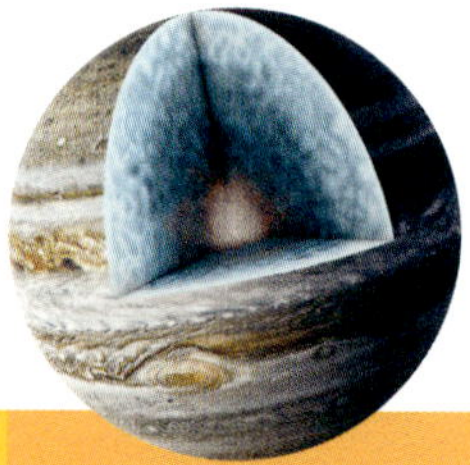

gas: a substance that is like air and has no fixed shape

core: the central part of a planet or other body

FACT

Winds on Jupiter reach 335 miles (539 km) per hour.

Far From the Sun

Jupiter is the fifth planet from the sun. It's an **outer planet**. Outer planets are cold. Jupiter is about -162 degrees Fahrenheit (-108 degrees Celsius).

outer planet: any of the solar system's planets that lie beyond the asteroid belt

Jupiter in the Solar System

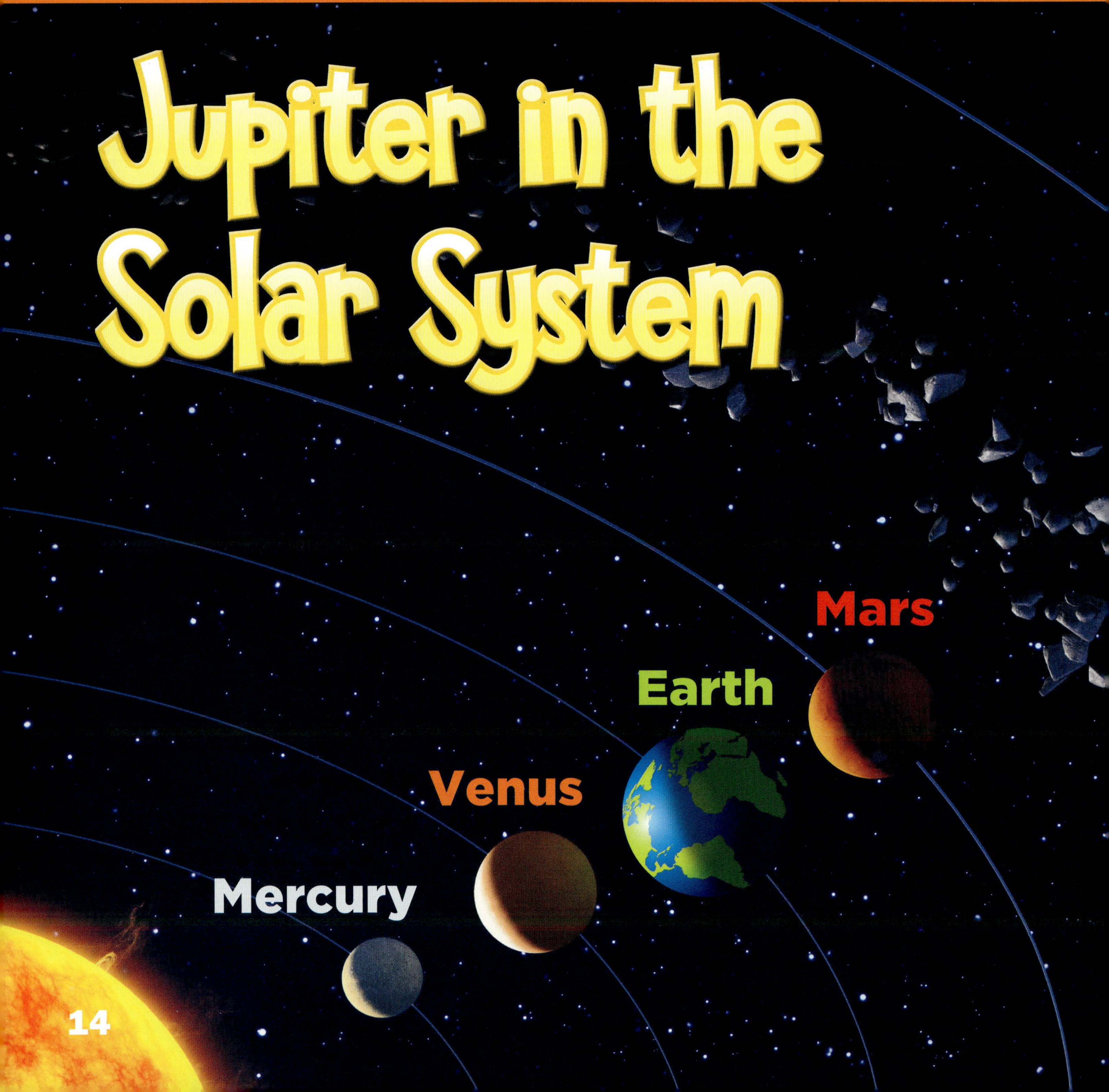

Neptune
Uranus
Saturn
Jupiter
asteroid belt

CHAPTER 3

Explore!

It can take six years to reach Jupiter. Only spacecraft have visited. They've taken pictures. Scientists studied these pictures.

FACT

A spacecraft discovered Jupiter's rings in 1979.

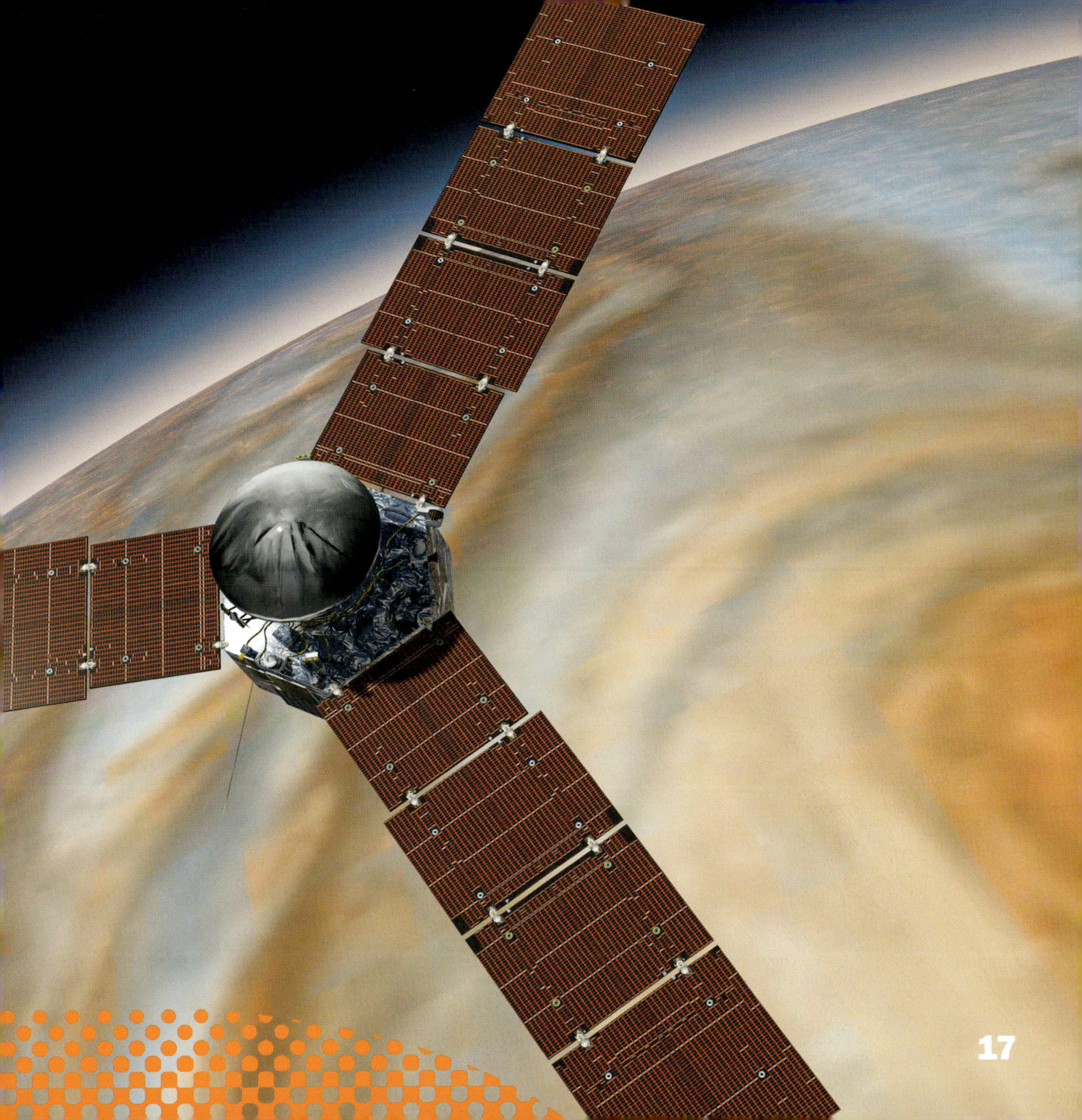

The Future

People keep learning about Jupiter. Another craft should reach it in 2030. It'll study the planet and its moons.

number of known moons

79 moons

Bonus Facts

All outer planets have rings.

Sunlight takes **43 minutes** to reach Jupiter.

Jupiter's days last about 10 hours.

Diamonds might **float** in the gas.

float: to be carried along by moving water or air

READ MORE/WEBSITES

Finan, Catherine C. *Planets and Moons.* Space. Minneapolis: Bearport Publishing, 2022.

Grack, Rachel. *Curious about Jupiter.* Curious About. Mankato, MN: Amicus/ Amicus Ink, 2022.

Milroy, Liz. *Explore Jupiter.* Planet Explorer. Minneapolis: Lerner Publications, 2021.

Jupiter Facts for Kids
nineplanets.org/kids/jupiter/

Mission to Jupiter
kids.nationalgeographic.com/space/article/mission-to-jupiter

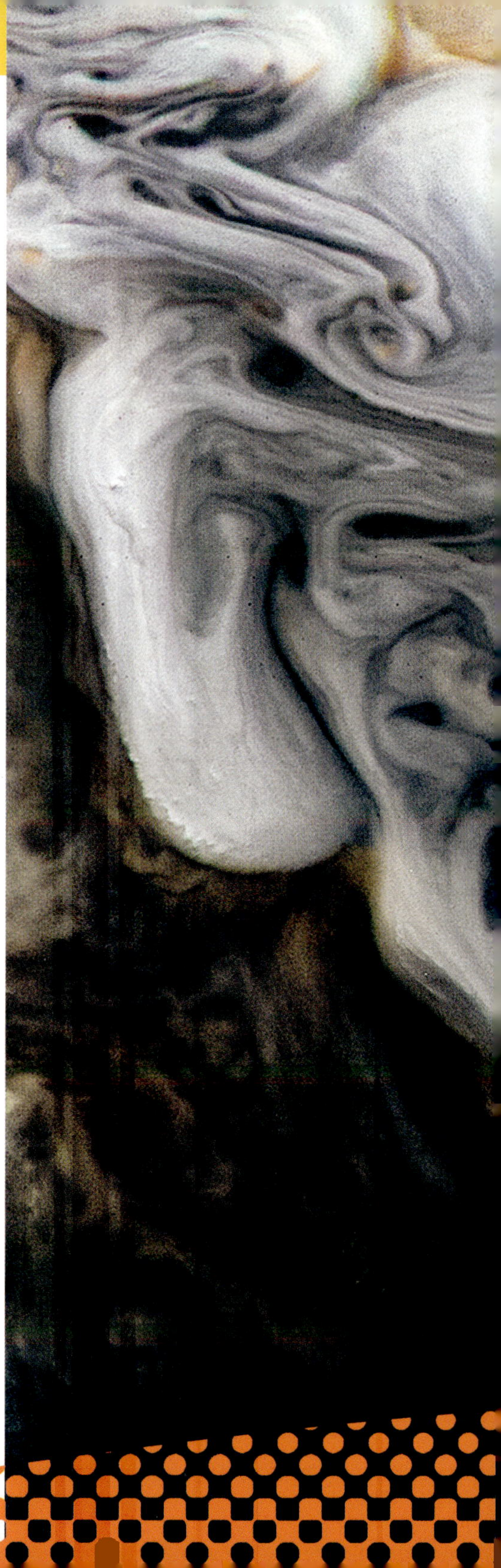

GLOSSARY

core (KOHR)—the central part of a planet or other body

float (FLOHT)—to be carried along by moving water or air

gas (GAS)—a substance that is like air and has no fixed shape

outer planet (OU-ter PLAN-it)—any of the solar system's planets that lie beyond the asteroid belt

rage (REYJ)—to continue out of control

INDEX